YOUR KNOWLEDGE HAS VALUE

AF148337

- - We will publish your bachelor's and master's thesis, essays and papers

- - Your own eBook and book - sold worldwide in all relevant shops

- - Earn money with each sale

Upload your text at www.GRIN.com and publish for free

Lea Weller BA

The Biological Adaptations of Camels in their Natural Environment

GRIN Verlag

Bibliografische Information der Deutschen Nationalbibliothek:

Die Deutsche Bibliothek verzeichnet diese Publikation in der Deutschen National-
bibliografie; detaillierte bibliografische Daten sind im Internet über http://dnb.d-
nb.de/ abrufbar.

Imprint:

Copyright © 2008 GRIN Verlag GmbH
Druck und Bindung: Books on Demand GmbH, Norderstedt Germany
ISBN: 978-3-656-54240-7

GRIN - Your knowledge has value

This essay will be describing the adaptations of a camel to its natural environment. It will include the camel's physical adaptations, their biological and physical adaptations towards heat and environmental temperature changes, water losses and gains. Also their tolerance to dehydration, and how they adjust their body's physiology to survive through it. It will explain how the camels hump actually works, what it stores and how it uses its stores for energy.

The camel's natural habitat is the desert. Deserts have extreme temperatures ranging from scorching hot summers to freezing cold winters. The Bactrian camel (two-humped camel) has a thick brown fur coat that changes through the year. It grows thick and dense in the cold winter to provide the camel with good insulation against the cold temperatures. In the desert, when it is summer the camel will shed its dense fur so that the camel can keep cool. The fur falls off in large sections very rapidly, leaving the camel looking almost naked. However, the Dromedary camel (one-humped camel) does not have any variations in its growth of fur. It is the same length all year round.

The camel has thick bare skin spots on its chest so it can lie on the hot desert sand. The thick patches raise the camel body slightly off the ground so it is not directly in contact with the radiation and heat from the ground. Camels have teeth so that they can eat thorny plants without doing any damage to the lining of their mouth. Camels can close their nostrils to protect the lining in their nasal passage, from blowing sands. Their ears are also lined with fur for the same reason of keeping sand out. To stop sand from entering the eyes of a camel they have long eyelashes for eye protection. The eyebrows of a camel are thick and stick out over the eyes to shade and protect their vision from the brightness of the day's sun. Their feet are flat, wide and have two toes that spread apart when walking so that the camel does not sink into the sand. The camels are herbivores and its usual diet consists of grasses, herbs, leaves and even thorns. It has thick lips so they do not feel any pain when eating the more thorny vegetation. Their usual fur colour is from light beige to dark brown, but can be white or almost black. This enables them blend in with their natural environment. They also have very good eyesight and a strong sense of smell.

The camel's main predator is the tiger and the wild wolf. The way in which they defend themselves is much like that of a horse. They kick their back legs and bite. The camel could also run away as it can run at speeds of approximately 45 miles per hour.

The camels hump (in the case of the Bactrian camel, humps) stores fat, which the camel metabolises for energy when there is a low food supply. The hump becomes smaller when the camel uses this fat store and it leans to one side or can no longer be visible. After feeding, the camels hump gets bigger, harder and returns to its previous size.

Unlike other animals the camel can change its body temperature through the day, from 34°C to 41.7^{o}C. This helps the camel retain water, as it is not sweating when the temperature rises. An advantageous process that camels possess is the process in which they can selectively cool their brain tissue. Selective brain cooling, adapts the brain temperature so that it is cooler than the camels body. The venous blood that is in the nasal region and in the facial skin is cooler than the arterial blood that is heading for the brain. The venous blood cools the arterial blood before it returns to the brain tissue. This process is to protect the camel's brain from being damaged by the heat. This also helps in saving water. (Michal Caputa, 2004).

Camels usually take in heat during the hot days, by drinking less water. This helps it reduce its water evaporation rate, to help reduce its daily water loss. By drinking more water the camel's daily water loss by evaporation increases. The way in which the camel stores heat is by raising its body temperature. The camels have a particular behavioural pattern during the day. If water was limited then the camel would lie in the same spot on the ground, with its legs underneath its body. This helps reduce their body surface area. The camel would only move in the same direction as the sun moves in order to reduce its heat gain. Another way of reducing their heat exposure is to lie together in groups very closely, to again reduce surface area exposure.

 The way in which camels evaporate water is partially by the use of sweat glands that are situated all over the camels body surface area. So when it is subjected to heat, the glands secrete sweat. The sweating is not visible as it evaporates off the skin under its fur coat. When the camel's food supply is green and has a high content of water, the camel's water intake is high and its urination rhythm is increased. In the winter the camel's water stores are not used for heat regulation, so the kidneys expel the excess water through urination. The amount of urine will depend on the food supply's water content, and how much food the camel eats. The camel empties its bladder regularly as its bladder is quite small.

A camel can go for weeks without drinking water and it loses one third of its body weight, but has no effects on the camel's state of health. When the camel does drink, its water intake can be up to 200 litres at once. No other mammal can drink this amount of water. Its cardiovascular system allows the camel to store large amounts of water. During dehydration the camels blood concentration increases and the plasma volume decreases. Plasma osmolity and sodium concentration also increase. (Ben Goumi et al, 1993).

In conclusion, this essay has described the physical, biological and physiological adaptations a camel has to make to be able to survive in its extreme natural environment. The camel is a mammal that can achieve adaptations that no other animal can achieve. For example how it changes its body temperature, depending on the environmental heat changes, and how it can drink a large amount of water to store in its body for regulation of the heat or cold it feels from the environmental temperature. Camels can go for a much longer time without a water supply, but they still get water from the plants that they eat. They can rehydrate quicker than any other mammal. They also have the ability of selective brain cooling, which is a great advantage for protection against heat. It has been thought by some people that the camels hump(s) stores water but in the essay it has shown the true purpose of the hump(s). Camel are physiologically astonishing animals, in how they adapt to their extreme environments.

Bibliography

Cameron Hatch (2005) The Difference Between One-Hump And Two-Hump Camels, The Hatch Report, http://www.thehatchreport.com/information/camel-one-two-hump.html [online] (Accessed 26/11/08)

Caputa M. (2004) Selective Brain Cooling: A Multiple Regulatory Mechanism, Journal of Thermal Biology, Elsevier Ltd, 29, 7-8 http://www.sciencedirect.com/science?_ob=ArticleURL&_udi=B6T94-4DDXMRG-7&_user=10&_rdoc=1&_fmt=&_orig=search&_sort=d&view=c&_acct=C000050221&_version=1&_urlVersion=0&_userid=10&md5=36818727f2632f8d489ae6812b4f5d4a (Accessed 28/11/08)

Conger C. (2008) How Long Can A Camel Go Without Water? HowStuffWorks.com. http://animals.howstuffworks.com/mammals/camel-go-without-water.htm, (Accessed 28/11/08)

Department of Veterinary Anatomy (2008) Characteristics Of Dorsal Lingual Papillae Of The Bactrian Camel (*Camelus bactrianus*), Anatomia, Histologia, Embryologia, 30, 3, http://www3.interscience.wiley.com/cgi-bin/fulltext/118998363/HTMLSTART (Accessed 28/11/08)

Elkhawad A.O.(*1992)* Selective Brain Cooling In Desert Animals: The Camel (*Camelus dromedarius*), Comparative Biochemistry and Physiology Part A: Physiology , *Volume 101, Issue 2, Pages 195-201,* http://www.sciencedirect.com/science?_ob=ArticleURL&_udi=B6T2P-4867S35-35&_user=7564510&_coverDate=12%2F31%2F1992&_alid=830904467&_rdoc=38&_fmt=high&_orig=search&_cdi=4924&_st=5&_docanchor=&_ct=63&_acct=C000054125&_version=1&_urlVersion=0&_userid=7564510&md5=5db0045975265bc6390b3dcfbff85aa5 (Accessed 25/11/08)

Goumi M.B. et al (1993) Hormonal Control Of Water And Sodium In Plasma And Urine Of Camels During Dehydration And Rehydration, General and Comparative Endocrinology, 89, http://www.sciencedirect.com/science?_ob=MImg&_imagekey=B6WG0-45PMPW1-4Y-1&_cdi=6808&_user=7564510&_orig=search&_coverDate=03%2F31%2F1993&_sk=999109996&view=c&wchp=dGLzVtb-zSkWb&md5=efaea11e5fb0dd81a7d6c21fac938c72&ie=/sdarticle.pdf (Accessed 27/11/08)

Goumi M. B. et al, (1996) Water Restriction And Bone Metabolism In Camels, Reproduction Nutrition Development , volume 36, pages 545-554, http://rnd.edpsciences.org/index.php?option=article&access=standard&Itemid=129&url=/articles/rnd/pdf/1996/05/RND_0926-5287_1996_36_5_ART0010.pdf (Accessed 25/11/08)

Huffman B. (2004) Bactrian Camel, Ultimate Ungulate http://www.ultimateungulate.com/Artiodactyla/Camelus_bactrianus.html (Accessed 21/11/2008)

Lincoln park zoo, (2001) Bactrian Camel, Lincoln Park Zoo http://www.lpzoo.org/animals/factsheet.php?contentID=196 (Accessed 21/11/2008)

Pastoret P. P. et al (*1998*), Immunology Of Camels And Llamas, Handbook of Vertebrate Immunology, Elsevier Inc.

http://www.sciencedirect.com/science?_ob=ArticleURL&_udi=B863R-4PB8TXK-

J&_user=7564510&_coverDate=07%2F12%2F2007&_alid=830880194&_rdoc=20&_fmt=hi

gh&_orig=search&_cdi=35529&_sort=d&_st=4&_docanchor=&_ct=65&_acct=C00005412

5&_version=1&_urlVersion=0&_userid=7564510&md5=282d7c708ace455ca6075346fb242

06a (Accessed 25/11/08)

Tulgat R. (1992) Status And Distribution Of Wild Bactrian Camels *Camelus bactrianus*

ferus, Biological Conservation, Elsevier Ltd. 62, 1,

http://www.sciencedirect.com/science?_ob=ArticleURL&_udi=B6V5X-48Y1NJX-

FM&_user=7564510&_coverDate=12%2F31%2F1992&_alid=832205923&_rdoc=1&_fmt=

high&_orig=search&_cdi=5798&_sort=d&_st=4&_docanchor=&_ct=1&_acct=C000054125

&_version=1&_urlVersion=0&_userid=7564510&md5=db377f9f06ed604a9dfe16eb19b2092

0 (Accessed 27/11/08)

Yagil R. (2003) Camel (*Camelus dromedaries*), Israel Journal Of Veterinary Medicine, 53,

http://www.isrvma.org/article/58_2-3.htm (Accessed 27/11/08)